I04808256

CONTENTS

IndiCypWat

/ɪndɪsɪpwæt/

A revolutionary way to understand
industrial engineering terminologies

Build a strong industrial engineering
foundation in less than one hour

Author: Mohammad Amin

Change

Let's neglect the introduction and dive into the book itself, this is your first lesson, change your stereotype of finding an introduction, start improving right away, start by one aspect, then your life radically, without hard introductions and worthless spent breaths, your breath counter is counting down now, you must know where these breaths are spent.

"Judge yourself before Allah (God) Judges you"

Omar Bin Alkattab

If you are waiting for an explanation of the book's name, you will find it out while reading it.

Stop now! Remember what you've just read and consider how this book will bridge industrial engineering terminologies to anyone's real-life routine. In the end, the definition will be quoted...that's how the entire book will go on:

Expression → Industrial term

"change your stereotype" → Cultural Change
"start improving right away, start by one aspect" →
Kaizen Blitz
"then your life radically" → Kaikaku
"reading it" → Activity

Definitions

Culture Change: A term used in public policy making that emphasizes the influence of cultural capital on individual and community behavior.

kaizen Blitz: (Kaizen Event) a concentrated effort, typically spanning three to five days, in which a team plans and implements a major process change or changes to quickly achieve a quantum improvement in performance. Participants generally represent various functions and perspectives and may include non-plant personnel.

Kaikaku: A radical change, during a limited time, of a production system. Often means that an entire business is changed radically, normally in the form of a project. Kaikaku is most often initiated by management, since the change as such and the result will significantly impact business. Sometimes call a kaizen blitz.

Activity: A basic element of work, or task that must be performed in order to complete a project. An activity occurs over a given period of time.

Writing Down Mistakes

As soon as you make a mistake, write it down and describe how you did pass it, this approach prevents you from repeating the same fault again.

Such an approach is done in industrial facilities by writing a document of the breakdowns that happened to machines while operating and how it was solved, in order to revise it when needed.

Expression → Industrial term

"mistake" → Breakdown
"write it down" → Documentation
"how you did pass it" → Action Plan
"this approach prevents" → Prevention Action
"a document of the breakdowns that happened" → Maintenance Checklist

Definitions

Breakdown: A specific type of failure, where an item of plant or equipment is completely unable to function

Documentation: Any communicable material that is used to describe, explain, or instruct regarding some attributes of an object, system or procedure, such as its parts, assembly, installation, maintenance and use

Action Plan: Describing how to move from the current state to the future state

Prevention Action: Action taken to eliminate the causes of a potential nonconformity defect or other undesirable situation in order to prevent occurrence

Maintenance Checklist: An itemized list of discrete maintenance tasks that have been prepared by the manufacturers of the asset and/or other subject matter experts such as consultants. Checklists are the basic building blocks of a maintenance program.

Dart

I used to get bad grades in exams even if I really understand the subject. I would always say: "No problem! marks don't evaluate the real me, I just want to understand the concept and my answer isn't that far from right one"

The first part of my saying was correct, but the second was not. Yes, marks are not crucial, while understanding alone is not adequate, your answer should also be correct. If you tried the pain of a needle placed just millimeters near your vein, then you will understand how important these millimeters are!

Expression → Industrial term

"isn't that far from right" → Accuracy
"how important these millimeters are" → Precision

<u>Definitions</u>
Accuracy: The closeness of an indication or reading of a measurement device to the actual value of the quantity being measured. Usually expressed as ± percent of full-scale output or reading.

Precision: Measure of exactness, possibly expressed in number of digits, for example, computed to the nearest millimeter

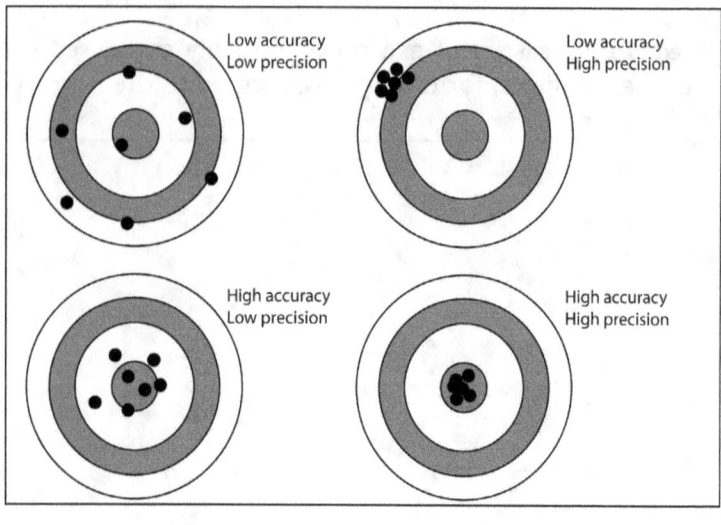

Thinking Out
Of The Box

Algazali - a Persian philosopher - said that if someone told you that he rubs two pieces of wood with each other and creates a small, red thing that eats the entire village, you would say he's crazy, but he isn't, it's fire! Thus, don't refuse ideas that -at the first glance- seems to be not in line with your beliefs.

ALWAYS think, think, and one more thing, think! Don't let any subtle detail passes by your eyes without questioning, why toenails lengthen slower than hand nails?

I always wondered, how to think out of the box? Until I read "think in a way that is not the way you used to, challenge your brain's way of thinking and finding solutions"

So there is always a way to your home that you don't know!

"Why do they call it rush hour when nothing moves?"

9

"you cannot read the label when you are inside the box."

Expression → Industrial term

"Don't let any subtle detail passes by your eyes without questioning" → Inspection

---

Definitions

Inspection: Any task undertaken to determine the condition of equipment, and/or to determine the tools, labor, materials, and equipment required to repair the item.

Heroes

In my opinion, heroes' purpose is to fill something that is lost in the community, so if there is no hero, we tend to create one; just like Superman who flies, and easily catches bad guys. If there is someone, like Martin Luther King, people mimic the best of them. People know heroes from their accomplishments and by telling each other about them.

Although heroes are imagined to be strong, but strength is not the point, or maybe it is! But what is strength? Is it defined by the number of kilos can be lifted? Or fighters have beaten? Or people have influenced? Well, it could be none of them.

> "The strong man is not one who is good at wrestling, but the strong man is one who controls himself in a fit of rage."

Prophet Mohammad -Peace be upon him-

Expression → Industrial term

"mimic" → Benchmark

"the best of them" → Best in Class
"telling each other" → Word of Mouth

<u>Definitions</u>

Benchmarking: Formal programs that compare a plant's practices and performance results against "best-in-class" competitors or against similar operations.

Best in class: A term for the top performing results for a particular metric. The term is used in benchmarking as firms typically compare results to the organization or industry that achieves superior results in a particular area.

Word of mouth: (or viva voce) is the passing of information from person to person using oral communication, which could be as simple as telling someone the time of day

Habits

Make pleasant habits to follow on a daily basis, even while making breakfast, reading, commuting to work, etc..., choose convenient places to sit regularly. By the time, you will get used to them, then you will feel peace there which leads to creativity.

> "*motivation is what gets you started. Habit is what keeps you going*"

Jim Rohn

Take a while thinking about habits you do regularly. Are they constructive or destructive? Do you have a plan to incrementally change bad one's? If not, write down its disadvantages and how it affects your life, is it worth it? If you can't get rid of it, consider learning about breaking bad habits.

Expression → Industrial term

"habits to follow" → Work Instructions
"making breakfast, reading, commuting to work, etc..." → Tasks
"incrementally change" → Kaizen

Definitions

Work Instructions: documents that clearly and precisely describe the correct way to perform certain tasks that may cause inconvenience or damage if not done in the established manner. That is, describe, dictate, or stipulate the steps that must be followed to correctly perform any specific activity or work

Task: One line on a task list that gives the inspector specific instruction to do one thing.

Kaizen: The systematic, organized improvement of processes by those who operate them, using straightforward methods of analysis. It is a "do-it-now" approach to continuous improvement.

Seeking
Knowledge
Overseas

Online learning is a grace, it opens a non-ending door of knowledge in front of you, in your mobile or laptop, it also gives you the ability to learn at the pace you like, revise anytime, and much more. Furthermore, it provides a numerous number of courses from different instructors and universities across the world, you are not forced to learn from a lecturer you do not like. Online learning reduces the cost for the lecturers' organization (such as the electricity bill; hence it is filmed once), and reduces the transportation cost for the student. It is a great opportunity to learn from different cultural perspectives and across seas.

"Live as if you were to die tomorrow. Learn as if you were to live forever"

Mahatma Gandhi

Expression → Industrial term

"electricity bill" → Fixed Cost
"transportation cost" → Variable Cost

Definitions

Fixed Cost: A cost that remains constant, in total, regardless of changes in the level of activity within the relevant range. If a fixed cost is expressed on a per unit basis, it varies inversely with the level of activity.

Variable Cost: A cost that varies, in total, in direct proportion to changes in the level of activity. A variable cost is constant per unit.

Trust Your Intuition

I believe that not everything should have an explanation, this does not mean that you should not think about answers, but means that the strong feeling of something should not be questioned too much, you should trust your subconscious mind.

In factories, there are negligible unavoidable issues that happen with no explanation! it is either from the machine or the operator, that cannot be avoided, unlike big issues that have a cause and must be avoided or solved.

Congratulations, you have made it up till here, now you can know what does the book title mean, its more than a couple of letters, it means your determination to dig until reaching an answer, it means your persistence to reach your goal! And the most important thing that you like the book!

IndiCypWat

Indi: industrial

Cyp: encyclopedia

Wat: are you waiting for an abbreviation? I told you, not everything should have an explanation, I have just trusted my intuition!

If you are still wondering why specifically these characters, why not "Xyz" for example instead of "Wat", ok, I will tell you. One day -before I started writing this book- I slept peacefully and dreamt that I am writing a book, and this was its name! If you are still skeptical, let me tell you more about sleeping:

- Donald Newman: dreamt about John Nash solving a problem that he couldn't solve
- Dmitri Mendeleev: saw the periodic table of elements in a dream
- Friedrich August Kekulè: found kekulè structure of benzene
- Beethoven, Paul McCartney, and Billy Joel: woke up humming music tracks
There are plenty of other examples...

Expression	→	Industrial term

"negligible unavoidable issues" → Chance Causes
"big issues that have a cause" → Variable Causes/ Assignable Causes
"avoided" → Preventive Maintenance
"solved" → Corrective Maintenance

Definitions

Chance Causes: Factors, generally many in number but each of relatively small importance, contributing to variation, which have not necessarily been identified. Note: Chance causes are sometimes referred to as common causes of variation

Assignable Cause: A cause factor designated as having a relationship to an accident. A causal factor or several causal factors assigned to an accident based on a predetermined causal classification system.

Preventive Maintenance: A maintenance process based on preventing unexpected events from occurring by employing proper maintenance procedures, clean environment, etc. Maintenance is mostly done during planned machine stops (fixed intervals). Emphasis is placed on replacing, overhauling, or remanufacturing an item at a fixed interval, regardless of its condition at the time. Scheduled restoration tasks and scheduled discard tasks are both examples of Preventive Maintenance tasks.

Corrective Maintenance: Any planned or unplanned maintenance activity required to correct a failure that has occurred or is in the process of occurring. This activity may consist of repair, restoration, or replacement of components.

Reading

Reading is like a money box, one day you may open it and find a treasure, do you remember your happiness when you opened it in your childhood? -assuming that you didn't spend all your money on chocolate!-. Don't stop reading if you couldn't get rapid results.

Please, don't read only for prestige, read to empower your brain and to put ideas and thoughts into your " ideas box".

When reading a book, start by defining what you want to achieve, then set duration to finish it. After finishing each chapter, revise it before starting the next one (chapter 1, then chapter 2...)

"You don't have to burn books to destroy a culture. Just get people to stop reading them."

Ray Bradbury

Expression → Industrial term

"money" → Resource

"reading a book" → Project

"defining what you want to achieve" → Scope Statement

"duration" → Constraint

"finishing each chapter" → Task

"each chapter" → A Phase

"chapter 1" → Predecessor for Chapter 2

"chapter 2" → Successor for Chapter 1

Definitions

Resource: Any person, groups, skill, equipment, or material used to accomplish a task, work package, or activity

Project: A temporary endeavor undertaken to create a unique product, service, or result.

Scope Statement: A definition of the end result or mission of a project. Scope statements typically include project objectives, deliverables, milestones, specifications, and limits and exclusions

Constraint: A restriction or a compelling force affecting freedom or action

Task: One line on a task list that gives the inspector specific instruction to do one thing.

A Phase: A time-based relationship between a periodic

function and a reference.

Predecessor Activity: Any activity that exists on a common path with the activity in question and occurs before the activity in question

Successor Activity: Any activity that exists on a common path with the activity in question and occurs after the activity in question

Practicing
As A Habit

Forming a new habit takes 66 days on average. Let's imagine that you formed this habit for reading books and started reading 40 pages a day -which will not exceed 45 minutes-, assuming that each book is 240 pages on average, this means you can finish a book every week with a free day! 4 books a month! 48 books a year! If you keep this habit for 20 years you will have read 960 books!!! This number is more than enough to make a profoundly wise, genius person, just by using a bit of your wasted time.

Force yourself to practice on a daily basis, doing so will shorten learning time for the long term, and give you a chance to get some free time. Otherwise, don't regret the knowledge you will lose.

When practicing for an assessment, calculate how much time is required to finish it - 2 days as an example -, it is crucial to have a day as a spare time just in case. Don't forget to buy a bottle of water from the cafeteria before it begins!

*"And when brought clear proofs, he (Jesus) said,
'I have come to you with wisdom and to make
clear to you some of that over which you differ,
so fear Allah and obey me'"*

Holy Qur'an [43:63]

Expression → Industrial term

"shorten learning time" → Learning Curve
"learning time" → Processing Time
"free time" → Idle Time
"how much time is required" → KPI
"2 days" → Lead Time
"a day" → Safety Stock
"just in case" → JIC
"buy a bottle of water" → Replenishment
"cafeteria" → Vendor

Definitions

Learning Curve: a concept that graphically depicts the relationship between the cost and output over a defined period of time, normally to represent the repetitive task of an employee or worker.

Processing Time: (PT) The elapsed time from the time the product enters a process until it leaves that process

Idle Time: Time during which a worker is not working

Key Performance Indicators: (KPI's) A select number of key measures that enable performance against targets to be monitored

Lead Time: The time that elapses from placement of an order until receipt of the order, including time for order transmittal, processing, preparation, and shipping

Safety Stock: (S.S) An extra quantity of a product which is stored in the warehouse to prevent an out-of-stock situation. It serves as insurance against fluctuations in demand.

Just in Case: (JIC) an inventory strategy where companies keep large inventories on hand. This type of inventory management strategy aims to minimize the probability that a product will sell out of stock.

Replenishment: the movement of inventory from upstream -- or reserve -- product storage locations to downstream -- or primary – storage, picking and shipment locations. The purpose of replenishment is to keep inventory flowing through the supply chain by maintaining efficient order and line-item fill rates.

Vendor: A general term used to describe any supplier of goods or services. A vendor sells products or services to another company or individual.

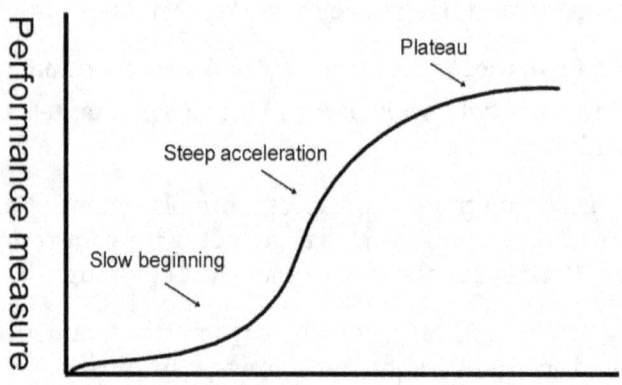

Number of trials or attempts at learning

Learning Curve

Parents

Parents commitment and love towards their children makes the children more successful and much happier. Also, agreeing on home rules makes everyone knows what is required from them, but sometimes too many rules may generate stress.

> *"From your parents you learn love and laughter and how to put one foot before the other. But when books are opened you discover that you have wings."*

Helen Hayes

Expression → Industrial term

"Parents" → Upper Management
"children" → Stakeholders
"successful" → Productivity
"much happier" → Loyal
"rules" → Work Instruction
"what is required" → Social Responsibility

"sometimes too many rules can generate stress" → law of unintended consequences

Definitions
Upper Management: includes individuals and teams that are responsible for making the primary decisions within a company
A Stakeholder: A party that has an interest in a company and can either affect or be affected by the business
Productivity: A ratio between the output of the wealth produced and the input of resources used in the process of any economic activity.
Loyal: Customer loyalty is an ongoing positive relationship between a customer and a business. It's what drives repeat purchases and prompts existing customers to choose your company over a competitor offering similar benefits.
Work Instruction: Sometimes called a procedure document, provides step-by-step instructions for completing a specific task.
Organizational Social Responsibility: The obligation of an organization to seek actions that protect and improve the welfare of society along with its own interests
law of unintended consequences: Taking action to improve something, only to make something else worse

Writing

Writing is a powerful tool for expressing ideas, it relieves and gives a feeling of satisfaction, do not say I cannot write, just hold the pen or pencil, and let your mind send its signals to your hand, holding the pencil is so romantic and inspiring. Create links and networks in your subconscious mind between writing and inspiration, good feelings, etc.., this is not only granted to writing, but any good habit also you like or feel will benefit you, you can rebuild your brain networks to accept it and feel delighted or any other feelings while doing it.

Do not seek perfection by rewriting the same paragraph over and over again, it may need some corrections, but not while you are having the inspiration. In most cases, you will write more than one paragraph without any mistake. It's important to discover them before another person reads your book and criticizes you, imagine sending all readers a new version for free, to restore your bright reputation as a sophisticated writer! Use spell-check functions to avoid spelling errors.

When we were kids, we used to write with a mechanical pencil, and to reduce the time needed to refill it back then, we would put an additional lead inside it, so when we needed to refill, we did it fast and easy. I was fascinated with those pencils, the process of disassembling and cleaning my pencil was so enjoyable for me. I assigned 5 minutes for cleaning it, and additional 5 minutes for defining whether there was a problem or not, and where.

"Science is a pray and writing is its leash."

Al-Shafi'i

Expression → Industrial term

"seek perfection" → Six Sigma
"rewriting the same paragraph" → Over Processing
"corrections" → Rework
"write more than one paragraph without any mistakes" → Rolled Throughput Yield
"discover them before another person reads" → Internal Failure
" before another person reads your book and criticizes you " → External Failure
"imagine sending all readers a new version for free" → Cost of Quality
"spell-check functions" → Poke Yoke
"an additional lead inside it" → POUS
"did it fast and easy" → SMED

"disassembling and cleaning my pencil " → TPM
"5 minutes" → Cycle Time
" defining whether there was a problem or not, and where. " → Measles Chart

Definitions

Six Sigma: A customer focused; well-defined problem-solving methodology supported by a handful of powerful analytical tools

Over-processing: Refers to doing more work, adding more components, or having more steps in a product or service than what is required by the customer.

Rework: The process of remedying an incorrectly manufactured part, sub-assembly, or assembly by reprocessing it either in the normal manufacturing system or in a dedicated facility.

Rolled throughput yield (RTY): The probability that a process with more than one step will produce a defect free unit.

Internal Failure: Scrap, Rework, Downtime, Yield losses Downgrading, etc....

External Failure: Complaint adjustment, Returned product/material, Warranty charges, Liability costs Indirect costs

Cost of Quality: Consists of the sum of those costs associated with: (a) cost of quality conformance, (b) cost of quality nonconformance, (c) cost of lost business advantage.

Cost of Quality Conformance: The cost associated with the quality management activities of appraisal, training, and prevention.

Cost of Quality Nonconformance: The cost associated with deviations involving rework and/or the provision of deliverables that are more than required

Poke-Yoke: A Japanese term meaning "mistake proof." A method of designing production or administrative processes which will, by their nature, prevent errors. This may involve designing fixtures which will not accept an improperly loaded part

POUS: (Point of use Storage) is storing materials as close to the location where they are used as is possible

SMED: (Single Minute Exchange Dies) techniques to reduce changeover times to the minimum possible. Focus not only on mechanical tasks but also on cleaning and on getting to specified conditions and properties quickly after the changeover

TPM: (Total Productive Maintenance) engages operators to improve equipment effectiveness with an emphasis on proactive and preventative maintenance.·Total: Total employee involvement·Productive: Eliminate or minimize

breakdown during production·Maintenance: Complete a preventive maintenance program

Cycle time: The total time one piece of product or one transaction resides in a process activity. It includes the setup time, process time, waiting for other units processed in the batch until the batch is released to the next process step

Measles Chart: A defect location check sheet (also known as a defect map) is a structured, prepared form for collecting and analyzing data that provides a visual image of the item being evaluated so that data can be collected visually rather than merely collecting a count of the number of defects

One Hand
Doesn't Clap

Working in groups always need to have synchronization among group members. The slowest member is the one who determine the work pace - *a chain is only as strong as its weakest link-*. I didn't mean that we have to tolerate an unacceptable amount of work; definitely there should be a baseline, but there is always "the slowest one". To keep the ship afloat, make sure to give them a new task before ending their previous one.

It is important to limit the number of tasks given for others; to make sure the team as a whole is consistent, and workload does not burden the slowest one. Group leader should always monitor -in place- producing the desired result as well as ensuring the least amount of wasted time.

Expression → Industrial term

"The slowest member is the one who determine the work pace " → Drum

"to give them a new task" → Push System
"before ending their previous one" → Buffer
"limit the number of tasks given for others" → Rope
"in place" → Gemba
"producing the desired result" → Effectiveness
"the least amount of wasted time" → Efficiency

Definitions

Drum, Buffer and Rope: (DPR) "The theory of constraints" method for scheduling and managing operations that have an internal constraint or capacity-constrained resource

Drum: The pace setting resource/Production pace (Constrain)

Buffer: The amount of protection in front of bottleneck resource

Rope: The scheduled staggered release of material to be in the line with the Drum's schedule

Push System: an inventory system where deliveries are planned in advance based on a master schedule

Gemba: (also written as Genba) is a Japanese word meaning "the actual place." In lean practices, the Gemba refers to "the place where value is created," such as the shop floor in manufacturing, the operating room in a hospital, the job site on a construction project, the kitchen of a restaurant, etc...

Effectiveness: The capability of producing a desired result or the ability to produce desired output. When something is deemed effective, it means it has an intended or expected outcome, or produces a deep, vivid impression

Efficiency: It signifies a peak level of performance that uses the least amount of inputs to achieve the highest amount of output. It minimizes the waste of resources such as physical materials, energy, and time while accomplishing the desired output

Efficiency and Effectiveness Matrix

		Low Effectiveness	High Effectiveness
Efficiency (Right Things)	**High Efficiency**	Low Effectiveness High Efficiency **Right Things** **Wrong Way**	High Effectiveness High Efficiency **Right Things** **Right Way**
	Low Efficiency	Low Effectiveness Low Efficiency **Wrong Things** **Wrong Way**	High Effectiveness Low Efficiency **Wrong Things** **Right Way**
		Low Effectiveness	High Effectiveness

Effectiveness (Right way)

At this point, we have finished storytelling, just pure engineering terms and concepts will follow.

5S

When you want to work, start by organizing your disk:

1- **Sort**: Remove all the clutter from the disk
2- **Set in order**: Organize in an efficient manner
3- **Shine**: Clean up the entire area
4- **Standardize**: Ensure standard ways of working for the first three stages
5- **Sustain**: Ensure that those stages are part of your culture

5S: "Typically attributed to the Toyota Production System (TPS), the overriding idea behind the Five Ss is that there is "a place for everything and everything goes in its place." Every item that is used in a business process is clearly labeled and easily accessible. Discipline, simplicity, pride, standardization and repeatability as emphasized in the Five S are critical to the lean enterprise in general and flow implementations specifically."

Brainstorming

Brainstorming is an incredible tool for generating ideas, you should strive for quantity not quality, encourage wild ideas, and do not evaluate them, evaluation time will come later. When brainstorming, you can use many ways depending on the situation and the members:

1- **Freewheeling**: Let everyone shout out ideas, and one or two of them take notes
2- **Round Robin**: Go around the table and take turns.
3- **Slip Method**: Let everyone write their ideas down and pass them to a scribe
4- **5 Whys**: There is a valuable way of thinking; to ask why, when first why is answered, ask the 2nd why, and so forth

Suppliers

S uppliers around any organization fall in one of four categories regarding how important they are:

1- **Strategic suppliers (critical)**: The ones with the highest impact on your work and with the highest risk if you lose them

2- **Bottleneck suppliers**: Suppliers that are rather small group, but they can significantly disrupt our business

3- **Leverage suppliers**: Suppliers provide organizations with regular things (stationery, water bottles, etc....)

4- **Noncritical suppliers (routine)**: Suppliers that are not important and have low impact on business and low risk

The Kraljic Matrix

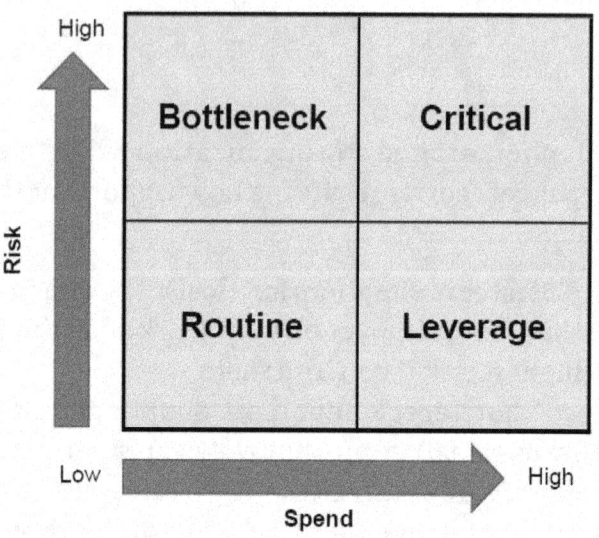

Demand

Every factory determines their manufacturing quantity depending on one of the following strategies:

1- **Level demand strategy**: Set up an average demand overtime, build inventory when they do not have demand, and draw from that inventory as they have more demand

2- **Chase strategy**: Ramp up production as they see increase in demand, decrease production as demand gets down

3- **Peak demand strategy**: They will have capacity for the highest possible level of demand

Conclusion

As any process, there should be an output, I hope this book had something to add to your knowledge. Moreover, to your personality.

I hope you can now see **industrial engineering** in your daily life, and see me in your way to distinction.

This book stops somewhere, but maybe this "somewhere" is the beginning of yours!

> *"We should be ashamed to die, until we've made some major contribution to humankind."*

Horace Mann

References

- Montgomery, D. (2013). Introduction to Statistical Quality Control (7th ed).
- Krajewski, L., Ritzman, L., & Malhotra, M. (2013). Operations Management Process and Supply Chains (10th ed.).
- Horngren, C., Datar, S., & Rajan, M. (2012). Cost Accounting: A Managerial Emphasis (15th ed.).
- Larson, E., & Gray, C. (2011). Project Management: The Managerial Process (5th ed.).
- Munro, R., Ramu, G., & Zrymiak, D. (2015). The Certified Six Sigma Green Belt Handbook (2ed ed.).